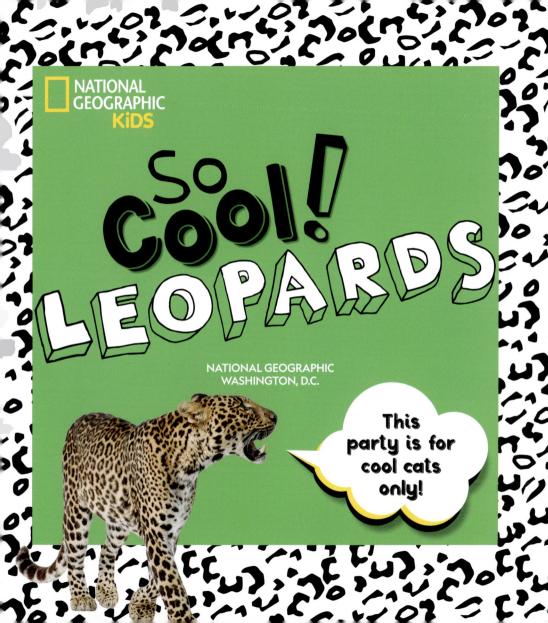

Can you spot
the spotted
cat sneaking
in the tree?

It's a leopard—a fearsome feline famous for its power, skills, and awesome spots.

What makes a leopard so special?

No two have the same spots!

They're super quiet.

They reach places other big cats can't.

Let's face it:
Leopards are ...

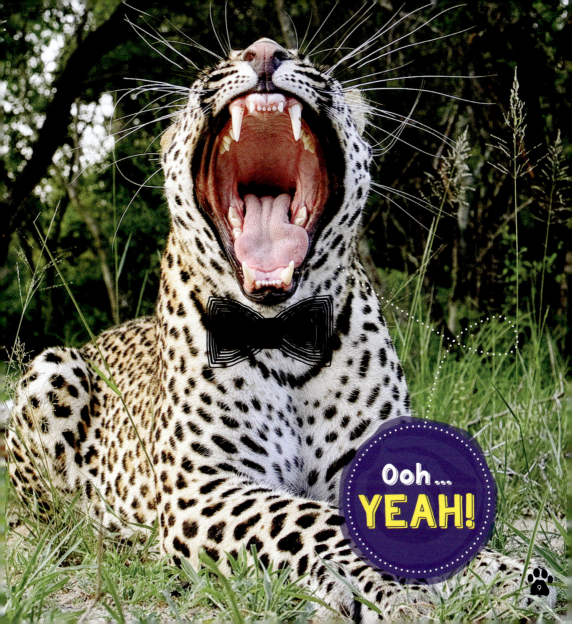

Leopards are the smallest of the "big cats," but they're super strong for their size.

THE BIG CAT CLUB

Your pet kitty can't be a "big cat" unless she can roar. Only leopards and these other cool cats can do that...

LIONS: Unlike all other cat species, these noble beasts live in groups—called prides—in Africa. They also have the strongest bites of the big cats.

Who you calling SMALL?

TIGERS: Tigers, the biggest of the big cats, can grow as long as 11 feet (3.3 m) from nose to tail. Their roars can sometimes be heard nearly two miles (3 km) away!

JAGUARS: These big cats in North and South America will travel up to 500 miles (804 km) to find mates and a home of their own.

SNOW LEOPARDS: Snow leopards have thick coats and fluffy tails, which they snuggle against for warmth in the chilly mountains of Central Asia.

SHHH! I'm playing hide-and-go-sneak!

Despite their might, leopards are supreme sneakers. Their spotted fur blends into grass and trees like an invisibility cloak!

SPOT THE DIFFERENCES

The leopard, jaguar, and cheetah all look alike from a distance thanks to their beautifully spotted fur. Here's how each cat's spots are special ...

LEOPARDS
This cat's spots—called rosettes because they're shaped like flowers—are made of brown circles outlined in black blobs.

JAGUARS
Slightly stockier than leopards, jaguars have a similar spot pattern with one key difference: Every rosette has a tiny black circle in the center.

CHEETAHS
This superfast cat is often a blur. Catch a cheetah sitting still and you'll see a coat covered with black dots rather than rosettes.

A stalking leopard creeps close to the ground, tip-toeing on the pads of its paws, patiently watching its prey, until, finally, it's close enough to ...

Just a little CLOSER...

Pounce! Leopards are built for leaping rather than speed.

LEAPING LEOPARDS!

Leopards are the superheroes of the cat world. They can jump over 18 feet (6 m), which would be like you leaping over three grown-ups lying head to toe on the ground. Scientists have also seen leopards tackling animals 12 times their weight. Imagine your mom stopping a car by jumping on it!

Leopards climb trees with the greatest of ease. They even wrap their tails around branches for balance!

CLIMB TIME!

Leopards are expert climbers. They scramble skyward to hide prey from scavengers or escape from lions. Researchers have seen leopards carry prey twice their weight up the trunks of trees two stories high. That would be like you climbing a ladder while carrying your dad or big brother … with your teeth!

Leopards have the most widespread habitat of all the big cats. They live throughout the southern half of Africa, across Asia, and in India and China. They live in jungles, grasslands, and mountains.

The secret to their success is they're not picky eaters. Some leopards have even mastered the art of catching fish!

LEOPARD MENU

Leopards will eat whatever animals are plentiful. Bugs, birds, fish, fruit—no meal is too small. They'll even hunt scary crocodiles!

Leopards are loving mothers. Babies will stick with Mom until they're about two years old, then head out into the world—where they like to live alone.

Leopards sometimes play by stalking animals too big to take down—even a giraffe!

Believe it or not, this black cat is a leopard. Leopards born with black fur are called black panthers.

BLACK CAT

The name "panther" can be confusing because it's also another name for the mountain lion, a completely separate species. But black panthers are 100 percent leopard, often born alongside other littermates. If you look closely, you'll even see that black leopards have spots—they're just very dark.

Leopards' excellent night vision lets them stalk around the clock. Instead of saying nighty-night to this leopard, we should say happy hunting!

Copyright © 2019 National Geographic Partners, LLC

Published by National Geographic Partners, LLC. All rights reserved. Reproduction of the whole or any part of the contents without written permission from the publisher is prohibited.

Since 1888, the National Geographic Society has funded more than 12,000 research, exploration, and preservation projects around the world. The Society receives funds from National Geographic Partners, LLC, funded in part by your purchase. A portion of the proceeds from this book supports this vital work. To learn more, visit natgeo.com/info.

NATIONAL GEOGRAPHIC and Yellow Border Design are trademarks of the National Geographic Society, used under license.

For more information, visit nationalgeographic.com, call 1-800-647-5463, or write to the following address:

National Geographic Partners
1145 17th Street N.W.
Washington, D.C. 20036-4688 U.S.A.

Visit us online at nationalgeographic.com/books

For librarians and teachers: ngchildrensbooks.org

More for kids from National Geographic: natgeokids.com

National Geographic Kids magazine inspires children to explore their world with fun yet educational articles on animals, science, nature, and more. Using fresh storytelling and amazing photography, *Nat Geo Kids* shows kids ages 6 to 14 the fascinating truth about the world—and why they should care.
kids.nationalgeographic.com/subscribe

For information about special discounts for bulk purchases, please contact National Geographic Books Special Sales: specialsales@natgeo.com

For rights or permissions inquiries, please contact National Geographic Books Subsidiary Rights: bookrights@natgeo.com

Written by Crispin Boyer
Designed by Julide Dengel

The publisher would like to thank everyone who worked to make this book come together: Ariane Szu-Tu, editor; Sarah J. Mock, senior photo editor; Joan Gossett, production editor; Anne LeongSon and Gus Tello, design production assistants; and Scott Vehstedt, fact-checker.

PHOTO CREDITS:
GI=Getty Images; NGIC=National Geographic Image Collection; SS=Shutterstock
Cover (UP RT), Life on white/Alamy Stock Photo; (LO RT), Eric Isselee/SS; (LO LE), Eric Isselee/SS; (UP LE), Eric Isselee/SS; back cover (LE), Paul & Paveena Mckenzie/GI; (RT), Eric Isselee/SS; 1, Eric Isselee/SS; 3, Michel & Christine Denis-Huot/Biosphoto; 5, Sergey Gorshkov/NPL/Minden Pictures; 6 (LE), Design Pics Inc/NGIC; 6 (RT), Andy Rouse/Nature Picture Library; 7, Valentin Wolf/GI; 9, Sergey Gorshkov/Minden Pictures; 10, ajlber/GI; 11 (UP), WLDavies/GI; 11 (tiger), Anan Kaewkhammul/SS; 11 (jaguar), Stephenmeese/Dreamstime; 11 (snow leopard), Jeannette Katzir Photog/SS; 12, Papilio/Alamy Stock Photo; 13 (UP), hphimagelibrary/GI; 13 (CTR), Waldemar Manfred Seehagen/SS; 13 (LO), Koilee/SS; 15, Vicki Jauron/Babylon and Beyond Photography/GI; 16, white whale/SS; 17, Martin Harvey/GI; 19, Inaki Relanzon/Nature Picture Library; 20, Havroshechka/SS; 21, Michael Durham/Minden Pictures; 22 (UP), Tribalium/SS; 22 (LE), corlaffra/SS; 22 (CTR), Kletr/SS; 22 (RT), Stéphan Bonneau/Biosphoto; 23, Ingo Arndt/GI; 24 (LE), Beverly Joubert/NGIC; 24 (RT), Robert Franklin Photography/SS; 25, Beverly Joubert/NGIC; 26 (LE), Paul Banton/Dreamstime; 26 (RT), Ann & Steve Toon/Nature Picture Library; 27, Buena Vista Images/GI; 28, irawansubingarphotography/GI; 29, Daniel Hernanz Ramo/GI; 30-31, Beverly Joubert/NGIC; 32, geraldb/SS

Hardcover ISBN: 978-1-4263-3525-9
Reinforced library binding ISBN: 978-1-4263-3526-6

Printed in China
19/PPS/1

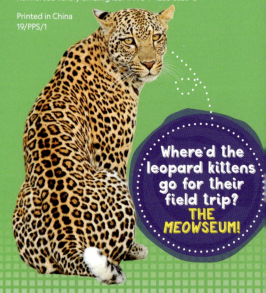

Where'd the leopard kittens go for their field trip?
THE MEOWSEUM!